HUMMINGBIRDS
FOR AMATEUR SCIENTISTS

By Aaron Brachfeld and Hank Braxtan

AUTHOR'S DEDICATION:
What is required to be known
May be learned through logic
Upon evidence and examples
Reaffirm and apply theory
To form a conclusion

ISBN-13: 978-1530469802

ISBN-10: 1530469805

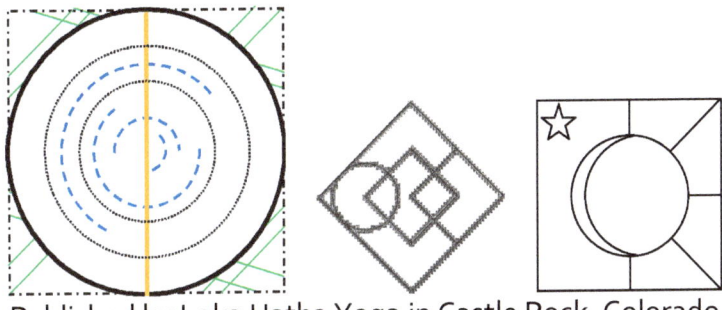

Published by Loka Hatha Yoga in Castle Rock, Colorado

There are many suitable places for your meditation.
We hope this becomes one of them.
Anguttara Nikaya 1.159

LOKAHATHAYOGA@GMAIL.COM - (719) 422-9536
http://lokareview.blogspot.com/p/books.html
lokahathayoga.blogspot.com

TABLE OF CONTENTS

Discovery ⸺ — —

"I just discovered this nest a few feet from our front door!"

What is Discovery? - - -

Discovery is learning "facts" for the first time.

What is a Fact? - - -

A fact is something that is true.

What is Science? - - -

Science is not a thing, but a process, or method. The method helps us understand why things are the way they are, and how things work. This process, or method, of science uses experimentation, observation, and logic.

There is a difference between "how?" and "why?" questions.

Questions that begin with "how" help us learn or discover "facts." The scientific processes we use to answer "how" are "experimentation" and "observation."

Questions that begin with "why" help us understand whether something is a fact and requires a process called "logic" or "reason." Logic is a mathematical process for understanding what is "true," or a fact.

SCIENTIST'S REFERENCE FILE:

- Hummingbirds will typically nest in places above the ground, sheltered and hidden from above and below.
- Though housecats are now the primary predator of hummingbirds, before people brought cats to America, other birds (especially hawks and jays) were the primary predators of hummingbirds.

HUMMINGBIRD SCIENCE: Why do hummingbirds nest there?

Pretend you are a hummingbird and a baseball is your nest. You'll want to put your nest where it is hidden from the sight of hawks and jays - whether they are hunting from the sky or the ground. We'll also see if these places are safe from cats who hunt in trees.

..._Fill in this data table below with checkmarks:_

	SEEN FROM SKY	SEEN FROM GROUND	SEEN FROM TREE BY CATS
TOP OF TREE:			
MIDDLE OF TREE:			
BASE OF TREE:			
IN TALL GRASS:			
IN SHORT GRASS:			

What does the data say? Where it is safest?

The places with the least checkmarks are safest... Is that where hummingbirds typically make nests?

FURTHER QUESTIONS:

- Are these nesting spots not effective against housecats? Think about how cats are unlike the natural predators of hummingbirds (other birds): they have excellent senses of smell.
- Do hummingbirds disguise the smell of their nests? Why haven't the hummingbirds learned to do this?
- Hummingbirds are threatened with extinction because we let our pet cats outside to play. What can we do about this?

UITZILOPOCHTLI —— — —

Before science, people used to answer the questions of how and why through "mythology," or an application of ethics and spiritual beliefs to natural phenomenon. People believed that all phenomenon were either "right" or "wrong," and the result of "good" or "evil" deities. These deities were expressed in symbolic terms of geographical features, weather patterns, or even animals – like hummingbirds. Using such a symbol allowed people to express the subtle and complex concept of their God's temperament toward doing what was right, wrong, or some combination of both right and wrong.

One of these symbols was a hummingbird. To the Mexicas, the hummingbird represented the God Uitzilopochtli. To them, he was the God of the sun and the God of war, and the protector of the City of Tenochtitlan (which today is called Mexico City).

He was worshiped 18 days of the year by a practice of ceremonial murder called by the Mexicas "Toxcatl." This ceremony required Priests kill human beings upon an altar to please their God under the belief that if they did not, the sun would not continue to shine over the earth – and, equally terrible, the Mexicas would suffer defeat in war. You see, war was very important to the Mexicas: they relied upon war to sustain their economy: defeat in war would mean they would have no food, no tools, no raw materials. This would be the end of their civilization!

Of course, such beliefs were not scientific – they were religious expressions of faith. And despite proper practice of Toxcatl, the Mexicas were ultimately defeated, and Tenochtitlan was destroyed (it was rebuilt as Mexico City). No amount of murder could change the fact that the sun is not a God, it is a kind of star around which the Earth orbits in space – and that victory in war is never certain.

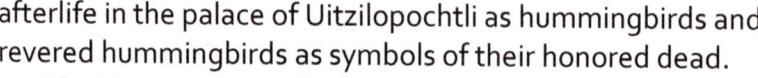

The Mexicas also believed that the warriors who died in battle would be reborn to an afterlife in the palace of Uitzilopochtli as hummingbirds and revered hummingbirds as symbols of their honored dead.

The Mexicas originally came from a place called Aztlan, and were called "Aztecs." They believed that Uitzilopochtli commanded them to abandon Aztlan and find a new home, and to stop calling themselves "Aztecs." He gave them a new name: the Meixca. Uitzilopochtli guided them through their journey, but for a time, they were left under the protection of Uitzilopochtli's sister, Malinalxochitl. The Mexicas resented her rule, and begged Uitzilopochtli to rescue them from her: Uitzilopochtli returned, put his sister to sleep and led them onward, leaving his sister alone in the desert. As you can imagine, Malinalxochitl was very angry. So, she gave birth to a son whom she called Copil to take revenge upon Uitzilopochtli.

Copil found Uitzilopochtli and the two Gods fought until

Uitzilopochtli tore out Copil's heart and threw it in the middle of Lake Texcoco, far, far away. Uitzilopochtli told the Mexicas to build a city when they saw an eagle perched on a cactus, eating a serpent. This was seen on Lake Texcoco where Copil's heart rested, and the Mexicas founded Tonochtitlan. The scene is remembered on the modern Mexican flag, which shows an eagle perched on a cactus eating a snake.

One story says that the Goddess Coatlicue became impregnated while sweeping a ball of feathers on Serpent Hill. Her other children (which included the 400 Centzonuitznaua and the Goddess Coyolxauhqui) were angered out of jealousy, and decided to kill their mother to prevent the birth of their new brother. At the moment of the murder, Uitzilopochtli burst forth from his mother's womb in full armor, fully grown, and defended his mother by chasing away his older brothers and sister.

In the age before science, this story explained why Uitzilopochtli, as the sun, appeared to chase away the stars (the Centzonuitznaua) and pursue the moon (Coatlicue). It also explained why the moon sometimes covered up the sun during an eclipse, and why the stars could be seen during that eclipse event. By presenting the sun as a kind of hummingbird, the Mexicas could better comprehend the reasoning of why the sun would be so ferocious.

Such beliefs mirrored the warlike culture of the Mexicas. Depending on war for their economy, they saw everything in terms

of battle and fighting – even the progress of the sun, moon and stars across the sky. In more peaceful cultures, the sun, moon and stars were not believed to be engaged in battle. Some of these more peaceful cultures even believed the sun, moon and stars to be friends, chasing each other in friendship or play. And in some other peaceful pre-scientific cultures, the sun, moon and stars were created by a friendly God to help people by giving light, heat and direction.

In our scientific age, we don't use mythology. We study the sun, moon and stars through the science of "Astronomy." We study their movements through the science of "Physics." And we have even constructed scientific implements to land upon the moon – and some of the "stars" which we now understand are other "planets" in our solar system. People have even stood upon the moon, to look back at earth – and bring back moon rocks for study!

Even though today few people believe that the sun is a hummingbird, or that the angry God of war is chasing his sister and brothers across the sky, the stories are important to learn if you will understand human nature – these stories are natural expressions of our species when confronted with things we do not understand.

The Mexicans are no longer warlike because their profound faith in the God of War was shaken. They adopted new scientific beliefs. Through "astronomy," Mexico learned that the sun was not a hummingbird, and they learned war is not a good way to sustain a civilization through a study of "economics." Over hundreds of years, the Mexicans have used science to learn new ways of living, adopting Democracy, and an economy based upon manufacturing, mining and agriculture – not war. Now, when they need food, materials or tools, they make them – and trade for them. And when they have disagreements with their neighbors, they use diplomacy to resolve the differences. Since adopting science hundreds of years ago, Mexico has remained a neutral party in all wars – except the Second World War, when Germany had attacked neutral Mexican ships, and Mexico was forced to defend herself.

Many famous scientists have come from Mexico. Most notably, in 1995, the Mexican chemist Mario J. Molina shared the Nobel Prize in Chemistry for work in atmospheric chemistry, particularly concerning the formation and decomposition of ozone. And Mexico is investing heavily in astronomic observatories – to learn more about our sun and other stars.

Observation ——— —— ——

"Mom is on the job."

What is Observation? - - -
Observation is the process by which facts and data are documented. Documentation is the process of communicating what is seen, heard, smelled, tasted, touched, or inferred. Documentation requires artistry, such as writing, drawing, photography, and other arts as well.

What is a Hypothesis? - - -
A Hypothesis is an understanding or a kind of premise on which we base our logical processes. Understandings can change with new facts and data.

What is Theory? - - -
A Theory is an understanding of how things work.

What is Data? - - -
"Data" are facts which are numerical or categorical in nature. For example, if it is a fact that you have 10 apples, there are two data: you have apples, and there are 10 of them.

SCIENTIST'S REFERENCE FILE:

- Male and female hummingbirds establish separate territories.
- Males will actually chase female hummingbirds out of their territory.
- Males and females must meet and mate in neutral territory.
- Hummingbird eggs require 14-23 days to incubate and hatch.

HUMMINGBIRD SCIENCE: What's a mother hummingbird's job?

So what is this mother hummingbird's job?

We observe right now she is trying to keep her eggs at the right temperature for incubation and hatching. But is this her only job? Will her job change? What will baby hummingbirds need when they are no longer in an egg?

Thinking about what changes and what stays the same will help us answer the question, "what is the job of a mother hummingbird?"

One big difference is that when they hatch, the baby hummingbirds need to eat. And since they belong to a species where dad doesn't raise the babies, we can "hypothesize" that mom will need to feed them.

Baby hummingbirds - like hummingbird eggs - need to be kept warm and safe, as well. And since with hummingbirds it is mom who does all the work, we "hypothesize" mom will protect and shelter the babies as well.

If we see this mother hummingbird do these things, we may confirm our hypotheses, and trust them as theories.

FURTHER QUESTIONS:

The "jobs" of mother and father are very different depending on the species. Some species need both a mother and a father to raise babies. Other need only a father - while others need only a mother. And in some species the babies raise themselves without mother or father! And to make things more complicated, within some species there are subspecies and even some individuals which have different behavior patterns than the rest of their species...

Hypothesize the job of human mothers - and the job of human fathers.

DEBATE —— — ——

Alejandro Rico-Guevara of the Department of Ecology and Evolutionary Biology at the University of Connecticut and Marcelo Araya-Salas of the Department of Biology, New Mexico State University were working together in studying hummingbirds when they discovered something different about Male Hummingbird beaks: there is a dagger-like structure at the tip!

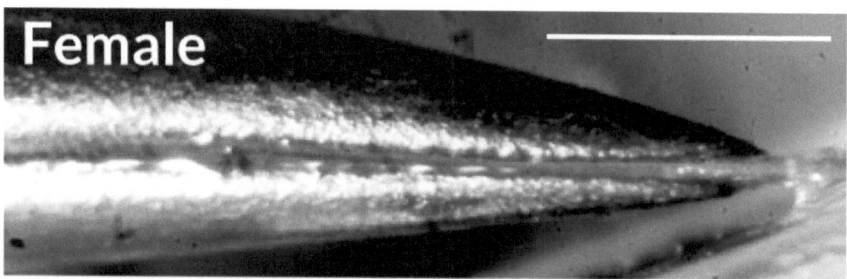

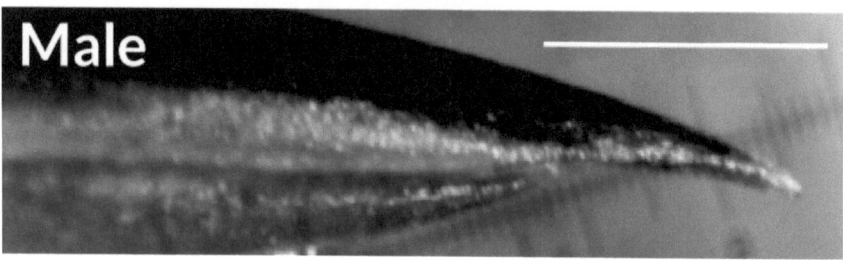

In a female long-billed hermit hummingbird (top), the upper and lower parts of the bill tip aren't very different. But in a male (bottom), the upper part grows longer and pointier in adulthood. The white scale bars represent 0.5 millimeters. Photograph by A. RICO-GUEVARA AND M. ARAYA-SALAS/BEHAV. ECOL. 2014

But this made them very curious – why would males and not females have this weapon?

To answer the question, they studied 5 leks (a "lek" is a place where animals demonstrate courtship behaviors) during 4 consecutive years using a combination of performance experiments, morphological analyses and behavioral observations. By this, they discovered that juvenile males acquired their dagger tips during the transition to adulthood, and that adult males with larger and sharper bill tips were more successful at dominating the lek – and therefore more successful in mating with females. The weapon was primarily used against other male hummingbirds!

The improved ability of the males with larger and sharper daggers to mate was resulting in new generations of males with even larger and sharper daggers – a case of sexual selection.

They quickly wrote up their observations in a story they titled, "Bills as daggers? A test for sexually dimorphic weapons in a lekking hummingbird." They submitted it to the Journal of Behavioral Ecology (a magazine published by Oxford University) in September of 2014 – and it was published the very next month!

Editors of scientific journals will check to make sure that the observations are true, that the methods used to collect them were sound, and that the conclusions developed from them are reasonable – by submitting them to other scientists who know about the subject. Then, by publishing the information, scientists (and the public) who are interested in the subject can learn about the interesting and consequential discoveries. And, even many years later, a scientist (or member of the public) who is interested in learning about something can research the subject.

By such research, effort is not duplicated: other scientists don't need to re-learn that adult male hummingbirds of this species have daggers, or that those daggers are used against other male hummingbirds. This allows scientists to more efficiently purpose themselves toward learning new things.

Articles are typically written in a way that scientists who are not familiar with the subject – or even the general public – might understand them. This permits broader application of what is learned: if a scientist who is an expert in another field of biology can understand hummingbirds use weapons to defend leks, they might be able to look for how males (or females) of other species use weapons for similar (or even different) purposes. Or if a scientist who is an expert in a non-biological field, say for example, a computer engineer interested in high speed photography, were to read the article, they might learn from the way that Rico-Guevara and Araya-Salasb used cameras to collect their data and contemplate an even better high-speed photographic system.

But sometimes, there are scientists who disagree. In such a case when there is disagreement, the disagreement is settled using debate methods. There is no "winner" or "loser" in the debate: the debate is held not to convince or win over the opponent. Rather, it is held for the benefit of the research: opponents will cite what they believe to be insufficiencies in the evidence provided so that the hypothesis might be strengthened, or a better understanding of the data developed.

After the research by Rico-Guevara and Araya-Salasb was published, Ethan Temeles of Amherst College in Massachusetts disagreed. The discovery contradicted a hypothesis of Temeles: Temeles believed that male and female hummingbirds had different beaks because they ate from different flowers – the sharper and pointier beaks of the males were simply required because of the shape of the flowers they ate from – not for fighting other hummingbirds.

Against this opposition, Rico-Guevara and Araya-Salasb presented their documented observations of the use of the bill as a weapon: numerous instances of bill-jabbing fights were seen. Rico-Guevara describes one intruder hovering over a male perched on a twig and darting in repeatedly to deliver eyeblink-quick stabs to the percher's throat.

Yet even that's not enough to convince evolutionary ecologist Ethan Temeles of Amherst College in Massachusetts. Temeles said he would only be convinced if he saw evidence that males with the most elongated bill tips prevail in fights because of those tips –

proof that winning fights with larger daggers earns males more females and offspring than males with smaller daggers.

Rico-Guevara then showed that males armed with larger and pointier bills were more likely than less equipped ones to win and hold a perch for courtship display - the proof required by Temeles. Temeles has maintained his opposition, but has not specified what would convince him. Temeles also did not provide any additional evidence to support his flower hypothesis, or contradict the dagger hypothesis.

Are you convinced by Rico-Guevara and Araya-Salasb?

Or by Temeles?

Science teaches that when we disagree about something, whether it is hummingbird beaks or even disputes arising from household chores – even matters of international policy – these disagreements can be settled through facts and logic.

Research

Baby hummingbirds!!!

What is Experiment? - - -

After observation is made, additional facts may be required to develop a hypothesis or theory. When all information is known, a result can be predicted: an experiment is designed to produce new observations required for more complete information. Factors which could affect data are systematically controlled and varied in an attempt to learn how and why things work through a process called "trial and error." When the variation in factors results in a predicted or desired result that confirms hypothesis or theory, the trial "succeeds." When the variation results with unexpected effect and yields undiscovered information, the trial "errs."

In example, new information that beak daggers are used for fighting shows an error in the previous theory. Errors are very useful in science!

What is Research? - - -

Research is the method by which data is systematically re-"searched" for undiscovered facts.

- Hummingbirds line their nests with spider silk and lichen. Spider silk and lichen are soft for baby birds, sanitary, and also can expand as baby birds grow bigger.
- Hummingbirds are among the world's smallest birds. They are so small that when collecting spider silk for nests, some hummingbirds get stuck in the webs like insects - and get eaten by the spiders!

HUMMINGBIRD SCIENCE: Biological interdependence

All species are interconnected in some way: every species helps some other creature, and harms some other creature, at least a little. Many people don't like spiders, but hummingbirds are interconnected with spiders – and numerous other animals and plants.

But what other species are interconnected with spiders? What other animals would be affected should spiders not be present in the environment?

Re-searching our observations for the facts is required. Take a look at the spider webs again and see if something new can be learned?

Find a spider web nearby your home - what kinds of insects has the spider caught? Obviously, the spider is interconnected with these insects. But do any of these insects have effects on other animals – or people? On hummingbirds? Are some helpful or harmful? For example, some mosquitoes and flies carry diseases – these would have a harmful effect on the animals affected by the diseases they carry. But other mosquitoes and flies pollinate flowers – these would have a helpful effect on the animals which rely on the plants pollinated. And what if animals were not harmed by the mosquitoes? Would their populations explode unsustainably to harm hummingbirds? Or people?

Make a list of all the plants and animals affected by the insects affected by the spider – and by this "web of life" see how important every living creature is to hummingbirds and each other.

FURTHER QUESTIONS: We know the fact that hummingbird directly relies on the spider for its nest – but how does the spider rely upon the hummingbird? Find the answer by re-searching the data you just collected.

FLUID DYNAMICS —— —— ——

"Wild hummingbirds got used to the bright lights and big cameras - ready to be our movie stars" - Kristiina Hurme, describing preparations made for the experiment: if the hummingbirds were affected by the lights, accurate data would not be collected! Instead of collecting data on the effect of flower shape on the eating habits of hummingbirds, they would be collecting data on the effect of light on the eating habits of hummingbirds!

Alejandro Rico-Guevara was encouraged by his inability to convince his fellow scientists of his theory and sought to provide further evidence that the shape of the beak of hummingbirds matters less to their ability to obtain food and more to their ability to defend leks. In this, he got help from his friends and fellow scientists, Margaret Rubega and Tai-Hsi Fan, both also of the University of Connecticut.

"Hummingbirds live life at incomprehensible speeds. Their

flight acrobatics are amazing, maneuvering more like insects than birds as they flit around, flying upside down and even backwards. They're a blur as they race between flowers. When they do pause to visit a flower momentarily, they're licking 15 to 20 times a second to extract their nectar fuel," explained Hurme.

It was this speed that first suggested to Hurme and Rico-Guevara that the existing hypothesis on how hummingbirds eat could not be correct. Rico-Guevera explained: "for over 180 years, scientists believed that to drink nectar, hummingbirds relied on capillary action. The idea was that their tongues would fill with nectar in the same way a small glass tube fills passively with water. The capillary action theory made sense since a hummingbird's tongue has two tube-like grooves. It would be a simple, passive way for nectar to travel up the tongue. But from watching hummingbirds in my native Colombia, we felt that capillarity just wasn't fast enough to keep up with how hummingbirds feed. We predicted that capillarity was too slow to account for the fast licking rates observed in free-living hummingbirds. Remember, they can drain a flower's nectar with around 15 licks in under a second! In our new study, we were able to slow them down on video to see how they really drink nectar. And what we found was quite different from the conventional wisdom since the 1800s."

Excitingly, their discovery resulted in a new understanding of fluid dynamics, and has promising applications for numerous disciplines of science. Their findings were published in August of 2015 by The Conversation, an online science journal purposed not only toward helping scientists learn about the work of other scientists, but helping the public access their exciting work, as well.

Here's an excerpt from their article, "Hummingbird tongues are tiny pumps that spring open to draw in nectar."

Four years ago, one of us (Rico-Guevara) and colleague Margaret Rubega challenged the conventional beliefs about capillary action for the first time. We showed that the forked tongue tips are not static, but dramatically spread inside the nectar, with fringed edges that open up like tiny hands. When the hummingbird retracts its tongue from the nectar, these fringes close due to the physical forces of surface tension and Laplace pressure, trapping nectar drops in their grips. Due to this transformation of the tongue shape, the tongue tips don't remain in the tube-shape necessary for capillary action.

We set out to study a medley of hummingbird species to see what these birds were really doing at the flowers. We needed a way to measure a tongue's thickness during the drinking process – straightforward, but not an easy task.

We designed see-through artificial flowers that we filmed with slow-motion cameras. From these videos, we could then track the shape of the tongue throughout the whole licking cycle. The difficult part was convincing wild hummingbirds to drink on command. Over time, we trained them by habituating them to the phony flower feeders and our whole filming setup.

When a hummingbird inserts its bill into a flower, it still needs to stick its long tongue deeper inside to get at the nectar within. After the tongue fills with nectar, the bird retracts the tongue back inside the bill. Researchers already knew that to keep the nectar inside the beak, the hummingbird squeezes the tongue with the bill tips as it is extended for the next lick. That compresses and flattens the tongue on its way out, leaving the nectar inside the bill. The way in which the nectar is moved from the bill tip to where it can be swallowed remains unknown.

To study the tongue-filling mechanism, we focused on the flattened shape of the tongue that each lick starts with. If the hummingbirds were using capillarity, once the nectar had made it into the bird's mouth, the tongue would immediately need to recover its tube-like shape before touching the nectar again.

By closely studying our slow motion videos of the birds drinking at the transparent flowers, we saw that the tongue remained flattened after the squeezing even as it traveled through the air to reach the nectar for another sip. It didn't snap back to its original pre-drink tube-like shape.

We studied 18 hummingbird species, and in hundreds of licks, we found that the tongue remained flattened until it touches the nectar. This was a key finding because it showed that the tongue didn't have the empty space inside needed for capillary action to work. Finally, we can confidently rule out capillarity as important for hummingbird drinking.

What we found goes beyond simply debunking capillarity. Hummingbirds have hit on an unexpected way to move liquid very quickly at this micro-scale: their tongues are elastic micropumps.

The grooves in the hummingbird tongue don't reach the throat, so the bird cannot use them as tiny straws. For this reason, instead of using vacuum to generate suction – imagine drinking lemonade out of a straw – the system works like a tiny pump, powered by the springiness of the tongue. The bird squashes the tongue flat, and when it springs open, this expansion rapidly pulls the nectar into the grooves in its tongue. It turns out it's elastic energy – potential mechanical energy stored by the flattening of the tongue – that lets hummingbirds collect nectar much faster than if they relied on capillarity.

While the tongue moves through the air, the elastic energy loaded into the groove walls during the flattening is conserved by a remaining layer of liquid inside the grooves acting as an adhesive. When the tongue touches the nectar, the supply of fluid allows the release of the elastic energy which expands the grooves and pulls the nectar to fill the tongue.

"As biologists, we were excited by this new discovery, but needed the help of an expert in fluid dynamics, Tai-Hsi Fan, to accurately explain the physics of this hummingbird micro-pump, and to make new predictions," said Hurme. "Our research shows how hummingbirds really drink, and provides the first mathematical tools to accurately model their energy intake. These discoveries will influence our understanding of their foraging decisions, ecology and coevolution with the plants they pollinate. Our ongoing research compares our new model with how much nectar hummingbirds drink at wildflowers, and looks at the trade-offs between drinking efficiently and fighting for dominance over territories either to attract females, to feed, or both."

While most data is easy to collect, data collection can be quite an adventure and some scientists have to travel to far and exotic places to get the data they need – requiring they construct elaborate laboratories and equipment. To develop this theory, Rico-Guevara and his friends had to camp in distant forests and also spend a lot of time in gardens they designed just for the purpose of data collection. Hurme accompanied her friend, colleague, and husband (Hurme and Rico-Guevara are a husband/wife team) to help.

Hurme explains that before any experiment, observation must be made to determine what additional data is required. "No matter where we are working, either in the forest or in a garden, we always start out by closely observing the birds' natural behavior. Luckily for us, hummingbirds have a lot of territorial behaviors and "routines" and so with careful observation we can figure out where best to put the cameras to film them feeding or returning to their favorite perch. If we are studying completely wild hummingbirds in a forest, then we usually cannot get close to them, and instead leave cameras recording the best flowers so that we can review the videos after. This requires many, many trips up and down the mountains to replace batteries on the cameras so we can film the birds all day long."

In designing garden experiments, these observations were applied. "As you know, hummingbirds readily habituate to gardens and feeders, so it is often very easy to study their behavior in semi- natural environments," Rico-Guevara said. "On our coffee farm outside of Bogotá Colombia, we have planted many hummingbird-pollinated flowers and put up feeders, so that we now have hundreds of hummingbirds flying around our farm every day. These wild birds have become accustomed to feeding on our different experimental feeders and very readily feed in front of cameras. This makes our fieldwork very easy because the birds basically come to us! But we still need to observe them using high-speed cameras, because they are just too fast for the human eye!"

Why do scientists like Rico-Guevara and Hurme study hummingbirds?

Rico-Guevara became fascinated with hummingbirds on his first field trip to the Amazon during university. While walking through the jungle, a hermit hummingbird zoom down and hovered in front of his face, literally checking him out before it sped away. "I remember being shocked that instead of being scared of me, this wild animal was actually coming up in my face to examine me, as if who was I walking in his jungle. It was that day I realized that hummingbirds were exceptional animals."

Hurme had studied the behavior of various monkeys, frogs, and birds before, but hummingbirds definitely "blew her away" as an exciting new challenge. "Their incredible speed makes them so impressive and the fact that we have to slow them down with cameras to really understand how they are interacting makes studying their behavior a never-ending challenge, but a fun one! I love how quick they are to learn new things. When we introduce a new type of feeder they learn almost immediately, and will never forget, and then continue searching for new things all around the coffee farm. It's really adorable to watch them taste and examine new brightly colored things on the farm, thinking that they might be new secret feeders!"

If you don't like to travel, Hurme says that you can still be a scientist. "Being a scientist doesn't require that you travel to far off exotic places - it just requires that you are aware and observing the world around you. In almost every place you go there are plants and animals interacting, behaving, surviving, you just have to pay attention to them and appreciate them. Once you start to realize that nature is all around you, and you focus on noticing the details, you can be a biologist anywhere! It can be really rewarding to notice the hawk that's flying overhead and think, maybe I'm the only person who really appreciating the beauty of this animal right now."

These scientists say that one of the most important lessons to be learned from studying living things is how they are all interconnected. "Hummingbirds are what they are because of drinking nectar. If you think about it, all their extreme behaviors that fascinate us, evolved to allow them to exploit floral nectar in the most efficient manner. So the differences in how every individual drinks nectar are the result of the same forces which have prompted the incredible diversification of hummingbirds - they feed on different flowers and through coevolution more plant and bird species can evolve by taking advantage of each other!

What this suggests for human behavior could have many interpretations, but we'd like to highlight the value of conservation and what everyone can do to help the wildlife around them. By planting flowers you can help hummingbirds and butterflies find food throughout their range. We never know what exciting discoveries we can make, and so it is important to help them continue to survive for the hummingbirds themselves and for the next exciting discovery awaiting future scientists!" said Hurme.

The second-most important lesson is that there is always more to learn. "The more we study hummingbirds, the more we realize there is more to study. Each discovery is a glimpse into their world that leaves us hungry to understand more. Using high speed cameras, we are able to slow them down and we are studying the details of their fights, calls, displays and interactions, in an attempt to understand what they are saying to each other. We are also interested in just how they can be so fast. We are in the process of training wild birds to different lights and feeders so that we might be able to measure how fast hummingbirds can see. We believe that they have "high-speed" vision! And we're excited to demonstrate that they are miniature real-live superheroes!"

Collaboration — ⎯ — —

"They're almost too big for their nest! Gotta fly soon..."

What is University? - - -
A university is a place where scientists study and work together collaboratively. These scientists typically also instruct and train students.

What is a Test (also known as a Trial)? - - -
A test is the process by which data to support a hypothesis is shown to be true, false, neither true nor false, or both true and false. And, if neither true nor false, under what conditions the hypothesis is true; and if both true and false, determine what elements are true and which are false.

What is Analysis? - - -
Analysis is the process by which elements of data are identified: categorical and numerical definitions are made based upon their ability to cause effects. Data which causes significant effect is understood to be relevant. Data which does not cause significant effect is understood to be irrelevant.

SCIENTIST'S REFERENCE FILE:

- Incubation of hummingbird eggs requires 14-23 days
- After hatching, hummingbirds will typically live for 5-10 years. The oldest hummingbird on record lived for 12 years.
- Some species can fly during their seasonal migration 500 miles without stopping - clear over the Gulf of Mexico.
- Hummingbird wings move faster than the human eye can perceive movement - to us, their wings appear to be a blur.
- Wings can serve many purposes - some birds use them to fly, others to swim, still others use them to help in walking.
- Hummingbirds use wings to fly, to make sounds for communication, to shelter baby hummingbirds and themselves, and for fighting.

HUMMINGBIRD SCIENCE: Library science

In studying hummingbirds it may become necessary to collaborate with other scientists. Not only do you need help collecting data, but other scientists may have better understandings of things that you don't.

For example, Rico-Guevara was unfamiliar with the physics of fluid dynamics. However, by working with other scientists, he was able to collaborate toward a more perfect understanding of the use of hummingbird beaks. Each of these scientists then gained new knowledge - not only of Rico-Guevara's biology, but in the science each one had specialized in: advances in fluid dynamics and ecology were made. Now, each scientist can bring their knowledge to bear in further collaboration with other scientists.

Collaboration can also happen through books, journals, videos, and other publications at libraries. Learning how to use libraries is essential for good science: if you are faced with a question you don't understand, or even understand how to understand, you can learn from others by reading, or watching recordings of them explain how they succeeded in a similar situation. Increasingly these journals and books are published on the internet.

Find Rico-Guevara's article in Journal of Behavioral Ecology for yourself. It can be as simple as asking your local Librarian (Librarians are a kind of scientist who study information).

TASTE SWEET? —— — —

Not all animals taste sweetness – or much of anything. But some animals carry a family of genes called T1R's. When an animal inherits T1R1 and T1R3 from its parents, it allows the animal to taste amino acids – what sophisticated human gourmands describe as "savory" or "umami" tastes. And when an animal inherits T1R2 and T1R3 from its parents, it can taste sugar – and then wants to taste more sugar.

Animals who can taste sugar will do almost anything to keep tasting sugar. A chicken cannot taste sugar, and given a choice between sugary sweets and seeds, will always pick the seeds – every time. This is because a chicken has not inherited the ability to taste sugar from its parents.

If you look at the parents of that chicken – the rooster and the hen – they could not taste sugar either. Neither could the roosters and hens that were their parents. Looking back hundreds, thousands, hundreds of thousands – millions and millions of generations of chickens, you would eventually see a kind of animal that is not a chicken. In fact, it would be classified as a dinosaur.

But how can a chicken be related to a dinosaur? The same way that a hummingbird became different than its parents.

When two animals mate sexually to combine their genetic material, sometimes an error occurs. This "mutation" results in some slight advantage (sometimes). Sometimes the mutation results in a disadvantage. When an animal has an advantage, it will be more fit – it will produce more children, it will be stronger and smarter.

For example, some male hummingbirds have sharper or longer daggers than others, and this gives them an advantage in fighting for females – and allows their male children to possess their same sharper or longer daggers. And if one of their male children has an even sharper or longer dagger, they would have an advantage over their father – and their brothers!

There lived an animal who was not quite a bird, and not quite a dinosaur - called Achaeopteryx. Archaeopteryx's parents were dinosaurs called Theropods.

Small Theropods related to Compsognathus (like Sinosauropteryx) probably evolved the first feathers. These short,

hair-like feathers grew on their heads, necks, and bodies. The feathers seem to have had different color patterns – just like in modern birds.

Through many generations, several new types of feathers developed. One was branched and downy, good for insulation. Others had a central stalk, with unstructured branches coming off it and its base – these were great for display or camouflage. Still others had a structure in which the barbs were well-organized and locked together by barbules – good for everything from flight to waterproofing.

Changes in the feet and wrists/ankles of these dinosaurs also occurred. The first Theropod dinosaurs had hands with small fifth and fourth digits and a long second digit. The fifth digit and then the fourth were completely lost. Then the wrist bones underlying the first and second digits consolidated and took on a semicircular form that allowed the hand to rotate sideways against the forearm. This eventually allowed birds' wing joints to move in a way that creates thrust for flight.

Eventually, over many generations, many of their bones were reduced and fused, which helped increase the efficiency of flight. Similarly, the bone walls became even thinner, and the feathers became longer and their vanes asymmetrical, also improving flight. The bony tail was reduced to a stump, and a spray of feathers at the tail eventually took on the function of improving stability and maneuverability. The wishbone, which was present in non-bird dinosaurs, became stronger and more elaborate, and the bones of the shoulder girdle evolved to connect to the breastbone, anchoring the flight apparatus of the forelimb. The breastbone itself became larger, and evolved a central keel along the midline of the breast which served to anchor the flight muscles. The arms evolved to be longer than the legs, as the main form of locomotion switched from running to flight, and teeth were lost repeatedly in various lineages of early birds.

New species not only develop by evolving new genetic material, but also through changes to existing genetic material – and by changes to how existing genetic material is used. All of these changes occur in similar ways to the changes seen in hummingbird daggers – through slow, gradual, natural selection.

When genes are not used, they go to "sleep" or become "dormant." They are "not expressed." They are still there – just not used. You see, DNA is not the only genetic material: many complex processes are involved in the use of portions of the DNA through combination for expression.

You have numerous genes just waiting to be expressed at the right moment: if you started to smoke tobacco, you would activate some of your genes to protect your lungs. Your children would inherit both the genes AND the activation sequence – even if they are not smoking. This means that they may suffer from asthma because of your smoking habit – even if you quit smoking after they are born. And this is why those genes are not expressed until after the smoke entered your lungs: there are few advantages to asthma, except to protect the lungs against smoke.

But, similarly, if your children do not smoke, the gene deactivate and become dormant again – your grandchildren would not have asthma.

Some genes take many generations to activate – and some take many generations to de-activate. Some can become active during your lifetime, and deactivate during your lifetime.

The activation of genes is used widely in plant breeding, when drought tolerance or other environmental adaptations are required: plants are exposed to drought conditions, and their children are more ready to tolerate drought conditions.

Chickens descended from birds which subsided on grains and meat – just like they do. But they also descended (much, much further back in time) from dinosaurs which ate a more varied diet. Specialization in grains and meat resulted in the genes required for tasting other things becoming either dormant or extinct. This gave an advantage to chickens (and other birds who could not taste sugar): birds who could not taste sugar ate a diet more rich in amino acids, and became stronger, faster, smarter, and more reproductive.

But was the gene lost? Or just dormant?

Maude Baldwin, of Harvard University, sought to find out whether the gene from the dinosaurs from which birds descended was still present and dormant, or extinct. For this, she undertook tests on hummingbirds because hummingbirds are one of the few birds which can taste sugar.

With the help of fellow scientist Yasuka Toda to answer her question, the scientists cloned taste receptors from chickens, hummingbirds and the hummingbird's closest relative, the chimney swift (which eats insects). They found that the T_1R_1 and T_1R_3 taste receptors in swifts and chickens responded only to amino acids – but that the same T_1R_1 and T_1R_3 receptors in hummingbird responded only to sugar!

Now, remember that in other animals, a combination of T_1R_2

and T1R3 is required to taste sugar. This showed that the ancient genes were not awakened, but rather that a mutation had occurred that permitted the amino acid receptors to perceive sugars.

Baldwin hypothesizes that "perhaps ancestral hummingbirds that lacked the sweet receptor frequented flowers to catch insects. On occasion they accidentally consumed some nectar. Small mutations in T1R1 and T1R3 would have allowed them to taste this sugary liquid, giving them access to a vital source of energy. This could have given nectar-sipping individuals the evolutionary upper hand compared to insect-eaters."

Her hypothesis is based upon the theory of Charles Darwin, who discovered how animals in new environments learn which foods are worth eating and which should be avoided: a sense of taste. "Real taste [in] the mouth, according to my theory must be acquired by certain foods being habitual – hence become hereditary," Darwin reasoned: in other words, when in a new environment, an animal will find what is edible or tastes good, and then become more and more adapted to find, eat and digest that food with greater and greater efficiency.

Baldwin's results show that Darwin was spot-on. As soon as a bird could taste sugary nectar, it began to eat it. And it got better at eating it. Now, there are birds who specialize in eating sugary nectar.

She published her discovery in the August 2014 issue of the journal, Science. She titled it, "Evolution of sweet taste perception in hummingbirds by transformation of the ancestral umami receptor." She summarized her work in an "abstract" (a summary) by writing "Sensory systems define an animal's capacity for perception and can evolve to promote survival in new environmental niches. We have uncovered a noncanonical mechanism for sweet taste perception that evolved in hummingbirds since their divergence from insectivorous swifts, their closest relatives. We observed the widespread absence in birds of an essential subunit (T1R2) of the only known vertebrate sweet receptor, raising questions about how specialized nectar feeders such as hummingbirds sense sugars. Receptor expression studies revealed that the ancestral umami receptor (the T1R1-T1R3 heterodimer) was repurposed in hummingbirds to function as a carbohydrate receptor. Furthermore, the molecular recognition properties of T1R1-T1R3 guided taste behavior in captive and wild hummingbirds. We propose that changing taste receptor function enabled hummingbirds to perceive and use nectar, facilitating the massive radiation of hummingbird species."

Baldwin's curiosity is perked, and she is filled with new questions. "Future research may focus on other nectar-eating birds such as sunbirds and lorikeets, and frugivores like tanagers, and whether they have undergone the same mutations as hummingbirds, or if a different mechanism explains their penchant for sugary foods."

Experiment —— —— ——

"It appears that flight is imminent. 4...3...2...1..."

What is a Library? - - -

A library is a collection of information designed for research, managed by an informational scientist called a "Librarian." Libraries also permit scientists who do not live or work near each other to collaborate by learning from each other. Libraries also permit scientists to collaborate through time – and death: books and journals remain in libraries forever, and long after a scientist dies, their learning can be accessed – even hundreds of years later – wherever and whenever it is needed.

SCIENTIST'S REFERENCE FILE:

- Hummingbirds eat plant nectar, insects and spiders.
- Nectar is eaten by slurping it with their tongue.
- Nectar is 55% sucrose, 24% glucose and 21% fructose.

HUMMINGBIRD SCIENCE: How to best attract hummingbirds?

Once these hummingbirds fly away, we'll want them to come back. Hummingbird feeders are the way to go! Who doesn't enjoy watching the birds? Feeders attract them to home windows with artificial nectar. This nectar is sold commercially, usually in "instant" form, requiring water be added to the sugar. Would a higher

concentration of sugar to the liquid result in more
hummingbirds coming to visit?

A simple experiment of varying the composition of
sugar and water in the artificial nectar will yield
sufficient observations to conclude the optimal recipe:
counting the number of hummingbirds that visit the
feeder when different amounts of sugar and water are
provided will indicate which recipe should be chosen if
the most hummingbird visits are desired.

For a week, set out seven feeders with different
recipes and tally the number of hummingbirds at each.

RECIPE	A	B	C	D	E	F	G
SUGAR	0%	10%	30%	50%	70%	90%	100%
WATER	100%	90%	70%	50%	30%	10%	0%
NUMBER OF HUMMINGBIRDS PER DAY							

FURTHER QUESTIONS:

1) Was there something special that happened during
   one of the days that would have interfered with
   your observations: maybe it was cold that day,
   and hummingbirds were less active? Maybe you had
   never hung a feeder before and the hummingbirds
   hadn't discovered it yet? Repeating this
   experiment may result in different results.
   Repetition is very important – doing an
   experiment over and over again ensures chance
   occurrences that could interfere with accurate
   results are avoided: repeat this experiment ten
   times and take the average, then compare the
   average to the original results.
   ➤ What special conditions resulted in the
      different results? What does this teach you
      about hummingbird habits?
2) Do you notice a trend in the data: is there an
   increase in the number of birds with an increase
   or decrease in sugar content? Does the number of
   birds decrease after an ideal composition is
   reached?
3) What else might affect hummingbird behavior? The
   color of the feeder or the sugar? The
   environment of the feeder? Flowers nearby?
   Design new experiments to test these hypotheses
   by varying the factors systematically.

INSECT-LIKE FLYING —— — ——

Lots of creatures fly – not only birds like hummingbirds, but some mammals (like bats and squirrels), lizards (like snakes), fish, insects, spiders, even some plants. Each creature has through countless generations, by a process of natural selection, developed similar bodies: they all have wings.

Though wings have taken many forms in many species – some covered in hair, others in feathers, others in scales, or bare skin; some supported by bone, others with softer tissues – all flying creatures have wings in common. And similar behaviors in using those wings: wings are held stiffly for coasting, sailing or soaring on air; wings are flapped, beaten or waved to change direction, rise or fall in altitude, or hover.

This is because the efficiency of proper aerodynamics gives creatures advantages by saving energy, providing better maneuverability, and greater range.

Besides developing similar structures for flying, and similar behaviors for flying, all flying creatures have developed their capacity for flight for similar reasons, too: by being able to fly, animals were able to obtain foods others of their species could not (some food sources are high up, or a far distance away over mountains, deserts or water) – and they could evade enemies that others of their species might not by fleeing out of reach into the air. Some even raise their children far out of reach in the treetops and mountains.

Yet even within the same type of animal, some have become better at coasting, others at soaring, others are good at hovering. Though they are both birds, a hawk is very good at coasting, but not very good at hovering – and hummingbird is very good at hovering, but not very good at coasting. And though they are both insects, mosquitoes are very good at hovering – while butterflies are excellent at coasting.

Douglas Warrick, of Oregon State University, teamed up with Brett Tobalske at the University of Portland and Don Powers at George Fox University to understand the differences in how flying animals hover. In studying many kinds of wings, they discovered something very important: hummingbird wings are 50% more effective in downstrokes than insect wings, which are equally effective in upstrokes and downstrokes. And this was because the insect wings were essentially flat, whereas bird wings were not.

Bird wings have bones, and insect wings do not. Because of this, bird wings must be shaped wider at one end. Also, birds rely on feathers to push air, whereas insects do not. Because of both these facts, bird wings are directional: feathers taper in a direction away from the wider part of the wing. Consequently, though both

insects and birds move their wings in figure-8 patterns while hovering, birds will be more effective than insects in downstrokes – but less effective in upstrokes.

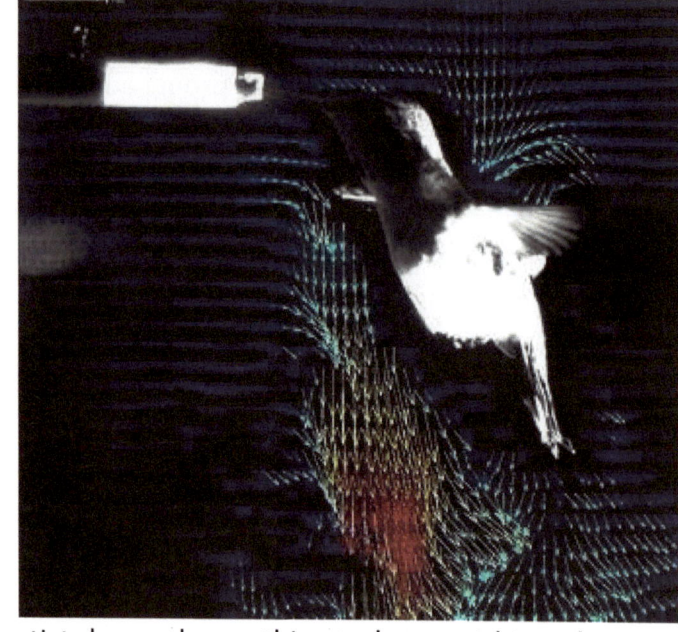

But the scientists learned something truly astonishing: the shape of the bird wing permits the hummingbird to use "leading edge vortices," an aerodynamic mechanism commonly taken advantage of by insects, to provide some of this lift on the downstroke: in effect, the more powerful downstroke of the birds is also more efficient.

"What the hummingbird has done is take the body and most of the limitations of the bird, but tweaked it a little and used some of the aerodynamic tricks of an insect to gain a hovering ability," explained Warrick. "They make use of what is, in other birds, an aerodynamically wasted upstroke. Coupled with taking advantage of leading edge vortices (which you can only produce to substantial effect if you're small) and voila, you're hovering for as long as you want."

This gives hummingbirds an evolutionary advantage, said Warrick: it permits them to eat from flowers what other birds cannot.

"It may not be the elegant, symmetrical flight of insects, but it works," said Warrick. "It's good enough. Hovering is expensive, more metabolically expensive than any other type of flight, but as insects have found, nectar from a flower is an even bigger payoff. Hummingbirds arrived at the ability to hover from a totally different evolutionary path. Natural selection made use of what materials were available (a bird body) and made a hovering machine."

Analysis —— — —

"Bye! I barely got this photo off. I missed the first baby's flight, but caught a bit of the second's this morning thanks to an alert cat sitting by the window!"

<u>HUMMINGBIRD SCIENCE: Attentive and ready to observe</u>

Lucky the cat was alert! It is sometimes helpful to have an assistant so you don't accidentally miss collecting important observations. But it is better still to analyze data to anticipate trends in activity.

Analysis is a process by which data is examined for un-looked-for information. It helps us understand what the data says - not just in terms of our experimental trial. For example, analysis of our experimental observations on hummingbird feeding habits teaches us not only the preferred sugar content of food, but that hummingbirds typically sleep at night, and are active during the day - and typically first fly 3 weeks after hatching. Knowing this, could you make sure to be extra attentive during the 18th through 24th day, all day long - and not miss making observations of the baby hummingbirds taking flight for the first time?

What kinds of scientists are there? - - -

There are several kinds of Scientists.

➢ "Amateurs" have no official training in science, but undertake science to the same standard as fully trained scientists. They are typically unaffiliated with Universities. Charles Darwin, Thomas Edison, Benjamin Franklin and numerous other famous scientists were Amateurs.

➢ "Masters" have official training in science, and have undergone the basic training of a "Bachelor" of science. They are affiliated with Universities. They gain their education and training in service to Doctors of Science, much like an apprentice in any other trade – as assistants. Sometimes in doing so, they contribute significantly to science. Michael Faraday, George Washington Carver, Alexander Bell, and numerous other famous scientists were Bachelors or Masters when they contributed some of their most significant work to science.

➢ "Doctors" have completed their official training in science, and have contributed in some significant way to science. They work and teach at Universities. The word "Doctor" means "to teach," and their primary function is as an educator – to train Masters.

➢ Some Doctors achieve so much that they are awarded special recognition: they become "Doctors of Science," "Doctors of Letters," or "Tenured." They have the right to study and teach at their University for as long as they live.

SCIENTIST'S REFERENCE FILE:
- Hummingbirds typically leave the nest 3 weeks after they hatch.
- Hummingbirds spend an average of 10-15% of their time feeding and 75-80% sitting and digesting.
- Hummingbirds will typically eat every 10-15 minutes, except at night when they enter into a state of hibernation.

FLYING MACHINE —— — —

Some people have studied birds, and wished they might fly too. When Leonardo da Vinci looked at insects, birds and other animals that flew, he understood that the reason they could fly was they were manipulating an invisible gaseous matter called "air." Though he could not see the material that the air was made of, he knew it was there: when it moved, such as by wind, or with a fan, or the wing of a bird, it carried force. And that force could be harnessed to let humanity fly just as well as any bird.

Leonardo da Vinci was not the first person to imagine such a thing. The ancient Han used hot air balloons in warfare for signaling, and the Incas flew in similar balloons above the deserts of Peru. And the ancient Greeks glided above the ocean on paragliders. However, da Vinci was the first person to imagine mechanical flight – a flying machine. He conceived of flying machines for soaring and hovering.

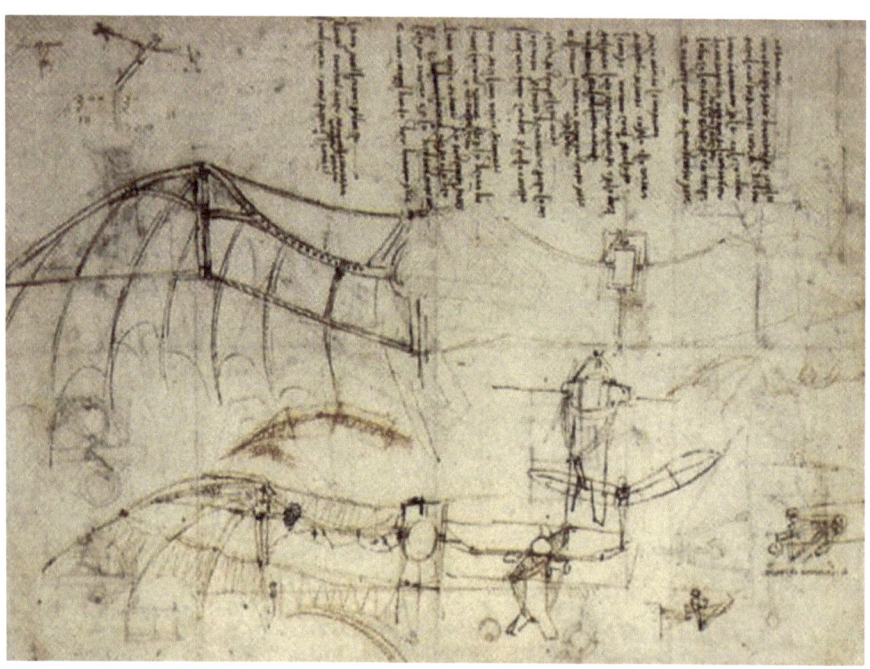

Mechanical flight is something which no other creature on earth is capable of doing – except humans.

When a new species evolves, they begin with what their parents gave them. When humans took to flying, we studied closely the anatomy of birds and insects, as well as their movements while in flight – this was something Da Vinci had learned to do when pursuing medical science.

Da Vinci was a good scientist – there was not a single subject he was not familiar with, his methods were efficient and logical. And he could communicate his understandings eloquently, through profound art and writing. Engineering uses for his knowledge to benefit his society, he transformed his world through technology.

Through science, he obtained enough of an understanding of how the flying animals of the world used motion to push air in the generation of force to replicate the movements with a machine. And then he Engineered that machine.

Just as other creatures took to the sky to obtain new food resources or for defense, so did humans: Da Vinci's intent was to provide something useful for commerce or war, the chief occupations of his culture. But his understanding of the advantages of flight was too far advanced for his age, and he failed to explain it. It would be hundreds of years before mechanical flight would be applied for either purpose by Wilbur and Orville Wright.

Today, mechanical flight is an indispensable part of modern commerce and war. We ship people, their goods, and their weapons at thousands of miles an hour through the air.

Though we were not born to fly, and did not evolve into a new species, the process was similar to biological evolution. We scientifically studied birds, insects and other flying animals to select their greatest successes and discard all their failures. Consequently, we can hover more precisely and longer with greater efficiency than any hummingbird, and soar better than any hawk.

Modern mechanical flight can be understood to have begun with the efforts of the Wright Brothers of South Carolina. Like Da

Vinci, the brothers were Amateur scientists. Neither graduated High School. Yet their mechanical flight was powered with greater energy than Da Vinci could have conceived of: in the interim, electrical and combustion engines had been invented by other scientists, and the Wrights adapted these engines for aircraft.

A new kind of flight was inspired: powered gliding. This was something no other creature had ever done before on our planet. Combining every advantage and discarding every disadvantage, the awesome form of flight inspired their American government. Their airplane was quickly put to purposes in commerce and war.

So many people saw the advantages of flight that soon Masters at Universities received funding from both commercial and military interests to improve that flying machine through formal science.

Quickly - within 60 years - mechanical flight became something entirely new. Evolution occurred again: now, there was spaceflight. For this, the electrical and mechanical engines were replaced with jet propulsion, nuclear fission power was then harnessed, and then nuclear fusion was implemented. Upon metal wings, we reached out and touched the moon, and the planets – and now we would reach ever further, to the stars themselves.

Our Universities continued to return investments into them with profitable science. And consequently, the modern Universities are looked to for similar successes in other industries: agriculture, medicine, chemistry – the value of knowledge was finally understood in the very practical terms we know today because of flight.

But this has come at a cost. We have as a society begun to lose faith in the Amateur scientist, and think that science only occurs at Universities.

While the science produced by our Universities is impressive, this should not be discouraging to an Amateur scientist. It is important to remember that it was because of amateur science that the professional scientist Buzz Aldrin would have never walked on the moon if the Amateur Wright brothers permitted humanity to achieve the full promise of the Amateur Da Vinci's mechanical flight.

Ethics ——— —— ——

"One of the babies flew last night, and the other this morning. A happy ending to what has been quite a journey. I still hear hummingbirds go whizzing and clicking by our window."

What is Ethics?
Ethics is the science of learning right from wrong. Right and wrong are abstract constructs used to describe how closely actions achieve goals for behavior: the ending of suffering is the primary goal for all living beings capable of experiencing pain, pleasure, strength, weakness, birth and death. Actions which result in the ending of suffering are "right," actions which result in suffering are "wrong." Actions which cause suffering are "wrong," actions which alleviate suffering are "right."

What is Non-Action (also known as No Harm, or Nonviolence)?
The practice of non-action (no harm, or nonviolence) is the theory resulting from scientific ethics: if suffering cannot be alleviated and action is required, the least or no harm should be caused. In other words, against a problem it may be better not to do something, or even to do nothing, than to do more harm than good.

SCIENTIST'S REFERENCE FILE:

- Hummingbirds possess a brain
- Hummingbirds by that brain feel pain and pleasure.
- Hummingbirds express emotion and complex thinking.
- Hummingbirds possess extraordinary memory.

HUMMINGBIRD SCIENCE:

The cat was alert enough to help him spot one of the hummingbirds as it flew by the window. Hummingbirds, too, are alert - just like cats or dogs, or other animals humans more closely associate with.

Possessing a brain, hummingbirds use their wings and voice to express emotions - even complex emotions like love and hatred. They exhibit complex thinking and problem solving, and possess extraordinary memory.

The theory that all creatures are self-aware and conscious has not been disproven: all creatures exhibit similar responses to pain and pleasure, and exhibit distress when sick, old or injured.

Understanding this, there exists a requirement upon all scientists to not unnecessarily cause distress or harm to animals which are studied - whether that animal is a hummingbird, an insect, or even another human being. Scientists must try to cause no harm through action - or nonaction. Of course, it is impossible to cause no harm at all, but scientists must cause as little harm as possible.

FURTHER QUESTIONS:

- ➤ How would observing hummingbirds potentially cause them to feel distress?
- ➤ Would proximity to the baby birds cause the mother hummingbird to feel fear?
- ➤ How do you think that harm can be minimized?
- ➤ Do people have an ethical obligation to relieve the suffering of other people, or even other kinds of life, like hummingbirds?

FEELING THE HEAT —— — —

A team of researchers with George Fox University in Oregon and the University of Montana has uncovered the ways in which calliope hummingbirds (Selasphorus calliope) get rid of the large amount of heat that is generated as they rapidly beat their wings (Donald R. Powers et al. Heat dissipation during hovering and forward flight in hummingbirds, Royal Society Open Science (2015)). In their paper published in the journal Royal Society Open Science, the team described their experiments and results – and a tremendous danger for birds in the future.

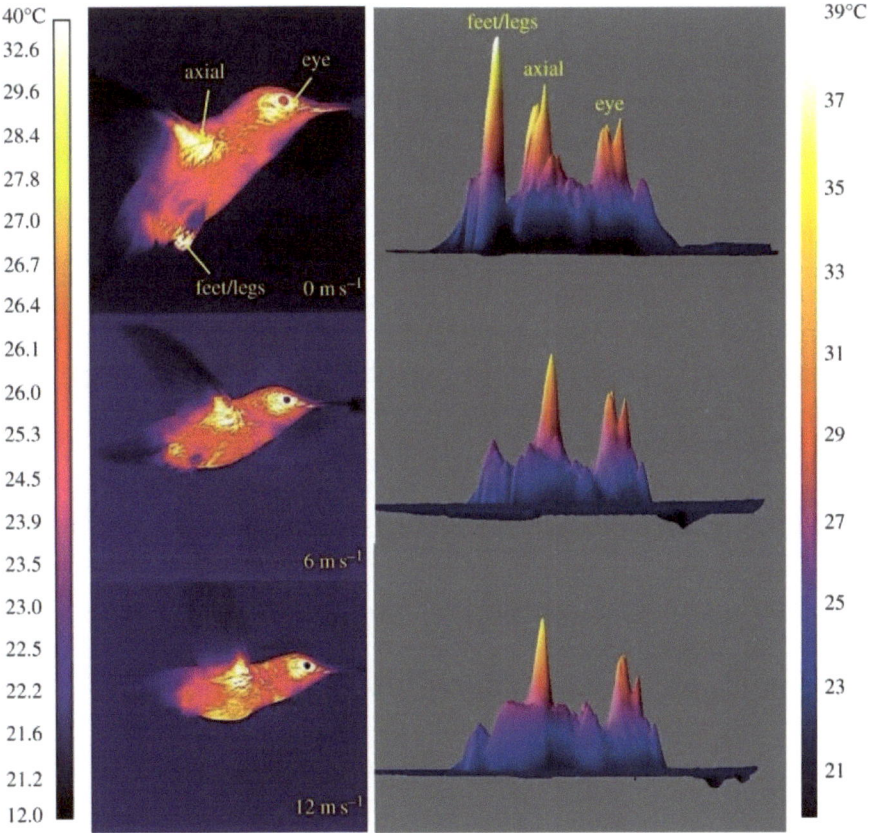

Whenever muscles are used, whether for flying – or walking – they burn calories and produce heat. We sometimes get so hot that we sweat (the evaporation of the liquid helps us cool off). But birds do not sweat, and are covered in feathers – making it difficult to keep cool when flying.

Filming hummingbirds in a wind tunnel with an infrared camera (which detects heat instead of visual light) and translating those infrared images into colors that can be seen, it was found that hummingbirds tended to dissipate heat from their feet and in areas where their wings met their bodies around their eyes: these were nearly 8°C (14.4°F) warmer than the rest of their bodies.

Interestingly, it was seen that when hummingbirds were flying slower or hovering, there was a greater temperature difference – the birds were cooling off when they slowed down and hovering. And at these times, the birds lowered their feet into the air to help them cool off even faster!

But as the world warms through global warming, hummingbirds (and other birds) may have a difficult time cooling down and could get heat sick, or die.

Victor Manuel Ortega-Jimenez and Robert Dudley from the University of California, working with the Smithsonian Tropical Research Institute, also are studying the effect of weather on hummingbirds. With more extreme weather expected as the planet warms, it is likely that hummingbirds will have to fly more frequently in heavy rain. Raindrops are not only like boulders falling from the sky for the tiny birds, able to cause injury upon impact, but can make the hummingbirds wet and have to carry extra weight. And hummingbirds need to eat so frequently that they cannot wait out a storm.

Some hummingbirds live in the tropical rain forests, where heavy rains are normal. Ortega-Jimenez and Dudley wanted to know how these hummingbirds coped with such dangerous conditions. So they put hummingbirds in a clear plastic box, doused them with varying degrees of simulated rain and filmed the whole thing with a high speed video camera. Their findings were published in the Proceedings of the Royal Society (Published online before print July 18, 2012): the birds cope mostly by changing their posture and working harder.

The conditions observed ranged from heavy, moderate, light and no rain. Under light or moderate rain, the birds did not even notice the rain at all: they flew in the same way as if it were dry. But with heavy rain, the hummingbirds had to increase their wing beats per second, and adjusted their bodies to a more horizontal stance. They also raised and lowered their wings less.

At first, it would seem a horizontal posture would mean more body area being bashed by the drops. But then, after watching the video over and over, it could be seen that in a horizontal position, the birds were better able to adjust the angle of attack of their wings, and doing so also tended to reduce the amount of rain that actually struck their wings. Changing position then, appears to give the birds more power while still expending as little as nine percent more energy. The end result is a tiny bird that is able to fly and feed in all but the most torrential downpours. It was calculated that "mechanical power output assuming perfect and zero storage of elastic energy was estimated to be about 9 and 57 per cent higher, respectively, compared with normal hovering."

What do you think? What will happen if hummingbirds have to regularly expend between 9% and 57% more energy because of more severe weather, and suffer heat stress when the weather is calmer?

What can be done?

Communication ⸺ — —

The soft nest has been mostly torn up. They would anchor themselves to the bottom to test their wings.

What is Applied Science?
Applied Science is the process of using logic and reason to solve problems through the discovery of new "facts."

What is Pure Science?
Pure Science is the process of experimentation or observation to discover new facts without an application directing the inquiry.

What is Communications? - - -
The process of sharing science with others is communication. Communication is typically undertaken through books, journals, newspapers and, increasingly, audio and motion pictures.

SCIENTIST'S REFERENCE FILE:
- This photograph shows the fact baby hummingbirds anchor themselves to their nest with their feet while testing out their new wings.

HUMMINGBIRD SCIENCE: What is the purpose of science?

Ethics requires the scientist not cause unnecessary harm or distress, whether through action or non-action. But what is the purpose of science? Why ask all these questions? Why bother learning anything at all?

Because the scientist knows that all beings experience distress when in pain, or when pleasure ceases, all beings experience distress when they are sick, old or dying, the scientist must do what they can to help. They direct their questioning with the purpose of alleviating distress and comforting harm – both their own suffering, and the suffering of others.

There are thousands of kinds of scientists – even scientists who study scientists! Some scientists act directly against suffering by disease and age through "applied science" – these scientists are called "physicians." Other scientists discover ways to increase the enjoyment of life through the "applied science" of "economics." Others use science to "engineer" new tools and technologies, buildings, airplanes, or other useful things we need to make life easier.

Most scientists, however, dedicate their efforts simply to discover new facts that can be used by "applied scientists": these are called "pure scientists," and practice "chemistry," "biology," "physics," or other "pure sciences."

Both applied science and pure science are equally important – and the skills learned in practicing them can be applied to every art and trade.

Studying hummingbirds is a practice of a very pure form of science called "biology" (the study of life): can you use or apply what you have learned?

FURTHER QUESTIONS: At the end of any scientific endeavor, it is important to take final observations and consider what new questions present themselves. What questions do you have now that you have learned so much about hummingbirds?

DRESSING THE PART —— — ——

Young hummingbirds look similar – it is difficult to tell a male from a female. Yet when mature, males are easily seen to have different coloration. This is not only because the female relies on camoflauge when sitting on eggs and caring for baby birds, these colors help adults recognize other adults, and in particular helps adult males show dominance to other males who might compete for the attention of the same females.

Animals (including humans) continuously demonstrate their social status to other members of their species using a variety of means. This important social behavior permits better organization and, ultimately, better fitness: by being able to perceive dominance readily, there is no need to continually spar and test dominance (continual fighting weakens those superior members of a species which would otherwise best thrive), and members of the opposite sex can immediately discern the best mating partners.

Shows of dominance are done using a variety of ways. Physical positioning (such as perching or standing taller than others) is common – as is making loud or complicated noises. But the most common method is the use of color.

Besides maintaining colorful feathers, some birds use other forms of color: putting beautiful stones, fabrics or metals in a nest site to attract the attentions of a mate and discourage competitors is very effective. While the preference for beauty is different in each species (a diamond is not a preferred stone for many birds, even if it is seen as beautiful by many humans), the preference for beauty itself has led to selections of mates who exhibit those traits seen as beautiful. All our ideas of beauty, virtue and goodness come from this process of evolutionary inheritance – since ancient times, this natural selection has resulted in countless creatures who appreciate beauty more and more.

For animals which do not wear clothes or make nests, beauty is an indicator of overall health (as feathers or fur usually lose their gloss and luster first) which would predetermine an ability to acqurie territory and defend it. For animals which use clothes (such as some species of crabs, humans, etc.) or make nests (as many birds do), access to desirable pretty fabrics, woods, stones, or metals indicates an ability to not only acquire those desireable objects, but to defend them as well.

It is the act of indicating a capacity for acquisition and defense which allows other members of the species to logically infer their relative capacity for acqustion and defense – and decide whether to challenge the dominance of their superior members. And it is the capacity to share that acquisition and defense which makes them worthy of mating: over many generations of selection, there is a tendency toward increased social behavior, toward generocity, self-sacrifice – prosperity and defense through cooperation. There is a tendency toward what we would describe as love.

Interestingly, Michael W Kraus and Wendy Mendes found that animals who proclaim a capacity for acquisition and defense, and experience no challenge to their dominance, experience physical changes to enhance their actual ability for acquisition and defense. Their fascinating research, published in the Journal of Experimental Psychology (Volume 143(6), December 2014 2330-2340: "Sartorial symbols of social class elicit class-consistent behavioral and physiological responses: A dyadic approach.") suggests that, to use a human expression, the clothes really do make the man.

Altering the clothing of men resulted in an alteration of their testosterone levels, and their ability to negotiate the price of a product. Those who dressed in clothes associated with high social status (a fancy suit) had higher testosterone and better negotiation success than those demonstrating low social status (a sweat suit) or random social status (participants wore their own clothes).

Quantitatively, their data showed that testosterone levels fell 20% in low status individuals, and high status clothing resulted in 200% more profit, and low status clothing resulted in 120% in losses: low status individuals were forfeiting gains to their social superiors, just as might be expected.

The loss of testosterone is similar as to what might be expected had the two men actually sparred – and the lower status man lost. In another study, the researchers found that testosterone levels in men drop after losing an athletic event.

Hajo Adam and Adam D. Galinsky in 2012 published in the Journal of Experimental Social Psychology a study on "Enclothed cognition." They found that clothes affect much more than testosterone – in highly social humans. In their work, they found that dressing the part can even affect technical performance: people were dressed in laboratory coats or in painters coats, and those dressed in laboratory coats had improved attention to detail in cognitive tasks.

Adam and Galinsky discovered something extraordinary and new: humans are the most social animals on Earth, and are also able to change their plumage. Hence, a new phenomeon could be described: "we introduce the term "enclothed cognition" to describe the systematic influence that clothes have on the wearer's psychological processes. We offer a potentially unifying framework to integrate past findings and capture the diverse impact that clothes can have on the wearer by proposing that enclothed cognition involves the co-occurrence of two independent factors—the symbolic meaning of the clothes and the physical experience of wearing them. As a first test of our enclothed cognition perspective, the current research explored the effects of wearing a lab coat. A pretest found that a lab coat is generally associated with attentiveness and carefulness. We therefore predicted that wearing a lab coat would increase performance on attention-related tasks. In Experiment 1, physically wearing a lab coat increased selective attention compared to not wearing a lab coat. In Experiments 2 and 3, wearing a lab coat described as a doctor's coat increased sustained attention compared to wearing a lab coat described as a painter's coat, and compared to simply seeing or even identifying with a lab coat described as a doctor's coat. Thus, the current research suggests a basic principle of enclothed cognition—it depends on both the symbolic meaning and the physical experience of wearing the clothes."

The "Cognitive Consequences of Formal Clothing" was also

observed by Michael L. Slepian, Simon N. Ferber, Joshua M. Gold, and Abraham M. Rutchick in March of 2015, and published in the journal of Social Psychological and Personality Science. They found "formal clothing enhances abstract cognitive processing. Five studies provided evidence supporting this hypothesis. Wearing more formal clothing was associated with higher action identification level (Study 1) and greater category inclusiveness (Study 2). Putting on formal clothing induced greater category inclusiveness (Study 3) and enhanced a global processing advantage (Study 4). The association between clothing formality and abstract processing was mediated by felt power (Study 5). The findings demonstrate that the nature of an everyday and ecologically valid experience, the clothing worn, influences cognition broadly, impacting the processing style that changes how objects, people, and events are construed." The clothing permitted better abstract thinking: in example, they more creatively assigned objects to categories (a camel is a form of transportation rather than simply an animal).

The understanding of the effect clothing has on identity has been intuitively understood long before it was scientifically described. Military uniforms and insignia are over symbols of social status, intended to intimidate inferiors. Sometimes, clothing is taken from the deceased or from superiors and worn to show respect, kinship, or adoption – a literal embodiment of their former social superiority. Some religions adopted clothing to imitate those of beggars or social inferiors to instill humiilty – while others adopted superior appearances to instill a sense of awe. When the Buddha Gotama instructed his students on the use of clothes, he said "Be content with any clothing, new or old, no matter how old, or what teacher it belonged to, what it was used for before, or whoever wore it before. Speak in praise of being content with any clothing. Do not, for the sake of clothing do any wrong, and not getting clothing do not become agitated. Getting clothing, do not become tied to it, attached to it, possess it. Uninfatuated, guiltless, seeing the drawbacks of attachment, understanding the escape from attachment, do not exalt yourself or disparage others upon the nature of their clothing. Use your clothing to remain skillful, energetic, alert and mindful. Do not, on account of this contentment, disparage others or exalt yourself" (Anguttara Nikaya 4.28).

Being highly social creatures, it is difficult to avoid social stratification and differentiation. In scientific communities, it is easy to try to say that amateurs and doctors are of either superior or inferior social status to each other. Yet, they are in different social groups: the Doctor practices science at a University, where they teach socially inferior scientists. But the Amateur works alone. Who is the inferior or superior of an Amateur?

Once this stratification occurs, however, other physical changes occur. When bonded, whether in marriage, or in the simple act of holding a child, a male's testosterone levels decrease to facilitate the sharing of the acquired and defended resources. Peter Ellison and Peter Gray (in the Harvard University Gazette, September 19, 2002) explained the advantage given to men whose testosterone levels would decrease upon bonding. "It makes sense, Lower levels of testosterone may increase the likelihood that men will stay home and care for their wives and kids, while decreasing the likelihood they will go out drinking with the guys and chase other women. Testosterone levels involve a trade-off between mating and parenting efforts. Single men invest only in mating, while fathers decrease their mating efforts in favor of parenting."

In Hummingbirds, the disadvantages of a single mother are clearly seen: it is difficult for her to care for the babies and herself, and when she must leave the nest, the nest is very vulnerable. Species which share parenting burdens have a distinct advantage – but species which do not also have an advantage: there is the opportunity for unbonded males to produce more babies.

Ultimately, the complex interaction between environment and species will dictate which path selection will take. For species in environments where many babies is a greater advantage (due to short life expectancy, intensive predation, or other pressures on young which could not be overcome by dual parenting), it will be uncommon to see the bonding required for dual parenting – or the propensity of male hormone levels to be affected by bonding. But where pressures are easily overcome by dual parenting, there will be an increasing tendency of females to select males as mates whose hormones are affected by bonding.

And where pressures are such that group parenting is either necessary or advantageous, babies will be seen to be raised by not just a mother and father, but more or less in an extended family – in a flock, troupe, congress, colony, or (in the case of humans) in towns and cities.

Consider your own town or city. Are most children raised by one parent, two parents, or extended families? What circumstances have led to this? What advantages – and disadvantages – do households with one, two or many parents have?

Now, as a human being, you have a unique opportunity no other animal does. Consider what choice you can make in raising your own family under these circumstances. Can you form a social network to give your children advantages and avoid disadvantages?

Consider yourself – do you possess high social status? Or low social status? Why? Unlike hummingbirds or any other animal, we have a choice in our social status. We can become more educated, we can become stronger in so many ways. We can acquire, defend – and share – resources, understanding that ultimately our best interests lie in the betterment of our community: through science, we can change our environment to produce less pressures upon us, so that we need not so constantly spar for scarce resources.

The Engineer ——— —— ——

A scientist must conclude from their work whether or not their hypothesis was correct – but this is not the end of their work.

This book on science must conclude with a description of the Engineer (sometimes called an "Inventor") – because that is the conclusion of all science. When we learn something, we must not only share it, but ethics requires we make it useful.

An Engineer, sometimes called an "Inventor," is the ultimate scientist, systematically and logically applying knowledge gained by their peers for the betterment of the world. Utilizing skills of business and craftsmanship, the Engineer can radically transform the way we live.

George Washington Carver was such an Engineer – developing peanuts from an experimental and theoretical crop into a mainstay of the American diet and a clean, renewable alternative fuel to petroleum. Thomas Edison was also such an Engineer – bringing light into the homes and businesses of every American. Faraday was a remarkable Engineer, giving to the world the gift of electrical mechanical power.

We so often are more familiar with Engineers than other scientists: perfecting science, it is proper we should be more familiar with them, because they have helped us in our own homes and lives. We all use a lightbulb, enjoy the benefits of industrial agriculture, and operate electrical engines daily. But we should not forget the invaluable contributions of their peers, without whom, they would not have improved our lives.

Without the information scientists at our libraries, without the books for our information scientists to find, how would we have ever discovered how to replace a broken heart – or any other disease our Medical Engineers have conquered? It was necessary to discover and understand the botany of the peanut before peanut oil or peanut butter be invented.

Imagine what hummingbird science will permit to be invented!

Imagination is the first step of invention.

Now, make what you imagined real – that is the work of an Engineer.

The application of logic and reason to solve our problems permits us to contemplate a world without the problems of today. It permits us to envision a better tomorrow – not only for ourselves, but every living creature. Yet such application relies upon the pure science of curiosity – they joy of learning.

Nothing is without cause. Nothing is without effect. The end of suffering is both possible, and necessary – whether you remain an amateur or become a professional scientist, or even an Engineer, there is much work to be done!

It has been in the hope that you would be encouraged to never stop learning, and the hope that in your learning, you would come to understand logic and reason, and thereby be able to imagine a better world, that this book was written and published. I hope that you will be a scientist.

Aaron Brachfeld, 2016
Castle Rock, Colorado

Become a scientist ——— —— ——

- If you are interested in research, contact your local university or college, and volunteer to assist the scientists there in their research. Learn all about data, its compilation and analysis.
- If you are interested in information, contact your local library, and volunteer the Librarians there. Learn how to store and find information.
- If you are interested in communication of that information, contact local publishers and journals, and volunteer. Learn how to seek out information to share, and how to best share that information. Learn how to write books and articles. Or contact the publisher of this book, Loka Hatha Yoga:
- Join or form a community of professional and amateur scientists – you can also contact Loka Hatha Yoga, if you don't have anyone to collaborate with in your neighborhood.
- In everything you do, remain aware of what things are, and try to understand how things work and why they do. Try to understand how to do what you love to do better. And if upon learning about the efforts of other scientists, you think you understand how to apply their science ethically, for the benefit of all beings, do not hesitate to Engineer a solution to the world's problems – good scientific ethics compels you to do all that is in your power for the sake of that love you have for the world that you study!

www.ingramcontent.com/pod-product-compliance
Lightning Source LLC
Chambersburg PA
CBHW040908180526
45159CB00010BA/2968